# Playground Math

Deanna Pecaski McLennan

Copyright © by Deanna Pecaski McLennan
First edition 2019

All rights reserved.

No part of this publication may be reproduced in any form, or by any means, electronic or mechanical, including photocopying, recording, or any information browsing, storage or retrieval system, without permission in writing from the author.

www.mrsmclennan.blogspot.ca

Joyful Math

# For my adventurer Caleb

There are many ways to play at the park.

Everyone has a favourite thing to do.

Running, jumping, climbing and swinging -

if we look closely, we will find math everywhere!

My bubbles float in the air.

I watch until they disappear.

Why do they change size the harder I blow on the wand?

I am building a house.

My trucks move boulders back and forth.

How can I find the heaviest rock?

The climber is so tall.

There are many ways to get to the top.

Can I measure how high and fast I climb?

Sometimes I ride my bike to the park.

My feet pump the pedals and the wheels spin around.

How will I know how fast and far I have ridden?

There is a tall tower made of tires.

So many steps to get to the top.

Do they each have a pattern?

The slide down is so much fun!

I close my eyes and pretend I am in space.

Why does the hole look smaller at the bottom than the top?

Balancing on the beam takes practice.

I walk in a pattern, one foot in front of the other.

How many steps can I take before I fall off?

The web is the most challenging of all.

I climb carefully, balancing on the ropes.

How many different ways can I move my body?

Jumping from the top makes me feel brave.

I crouch down before taking a giant leap.

Can I measure how high I sprang?

Visiting the splash pad can be refreshing.

The water bubbles up in a steady stream.

Why can I see my reflection in the water?

The slide is steep and fast.

The ride down makes me shriek and laugh.

Why do my shoes sometimes slow me down?

I grab on to the low tree limbs and slowly climb up.

Each time I make it a little higher.

How will I know if the branches can support my weight?

The climbing wall can be so tricky.

I have to use my hands and feet to hold on.

How can such narrow ledges hold my body?

Forwards and backwards I sway.

I love to fly as high as I can!

When my mom stops pushing, why do I still swing?

Listen to my beautiful music!

When I push the keys they each make a sound.

How many different patterns can I compose?

We see, hear, and feel math each time we visit the playground.

How many different ways does math help you play?

# Where's the Math?

As a mom I have always loved visiting the park with my three children. Over the years I have watched them explore the equipment and realized the rich, playful math that has emerged from their interactions with each other and use of the space.

Running, jumping, climbing, swinging, sliding, and balancing are activities that have embedded math. There is so much potential for helping children recognize, describe, analyze, and discuss these concepts. Some math prompts include:

**How do children move while at the park?** (e.g., running, walking, springing, sliding, balancing)

**Can children describe their movements?** (e.g., fast, slow, high, steady)

**How do children measure and track their physical movements while at the park?** (e.g., counting ladder rungs as they climb, marking their landing place after jumping)

**How do children's movements affect equipment?** (e.g., if they pump their legs harder, the swing moves faster and higher)

**What math can children see and describe in the playground?** (e.g., shapes, sizes, numbers, colours, patterns, spatial reasoning and logic, measurement)

**How does math affect playground toys and equipment?** (e.g., a swing moves like a pendulum, slides are angled, shoe friction on a slide slows down the rider, bubble sizes are affected by the amount of air blown into a wand)

**Why is math important for adults to know and use when designing playground equipment and layout?** (e.g., swings need to have adequate clearance, slides need to have safe angles, ropes need to be strong enough to hold a person's weight, safe but challenging ledges are on climbing walls).

**What extensions can be implemented to continue to support children's observations and questions, leading to mathematical inquiries?** (e.g., What happens if the angle of a slide changes? How does a pendulum work? Can you design and build a model of an amazing playground ride?)

I have written this book to help educators and families delve more deeply into considering the math that naturally happens at the playground. I hope that the photos inspire mathematical conversations as readers examine what is happening in each one, make rich connections to their own lives, consider alternative points of view, and think deeply about how math affects their playful experiences.

Math can be an authentic and engaging subject. Math happens everywhere, especially the playground. Next time you are there see what math you can find!

When we look at life through a mathematical lens, anything is possible!

~Deanna

Deanna Pecaski McLennan, Ph.D., is an elementary educator in Ontario, Canada. Deanna is fascinated by math and loves exploring its natural and authentic application in the living world. She hopes to help children and families recognize math as a beautiful and fascinating subject, and grow children's confidence, accuracy and interest in math.

Follow Deanna on Twitter and Instagram for regular updates including ideas for engaging children in playful, emergent math inside the classroom and beyond. Extending math learning outdoors is a favourite exploration!

Connect with Deanna:

deannapecaskimclennan@gmail.com
@McLennan1977
www.mrsmclennan.blogspot.ca

# Also from Deanna

Joyful
  Math

www.ingramcontent.com/pod-product-compliance
Lightning Source LLC
Chambersburg PA
CBHW040414220526
45473CB00004B/1236